KT-134-785

SCIENCE TOPICS

THE Earth & Beyond

CHRIS OXLADE

First published in Great Britain by Heinemann Library,
Halley Court, Jordan Hill, Oxford OX2 8EJ,
a division of Reed Educational and Professional Publishing Ltd.

Heinemann is a registered trademark of Reed Educational &
Professional Publishing Limited.

OXFORD MELBOURNE AUCKLAND
IBADAN JOHANNESBURG GABORONE BLANTYRE
PORTSMOUTH (NH) USA CHICAGO

Designed by AMR
Illustrations by Art Construction
Printed in Hong Kong

02 01 00 99 98
10 9 8 7 6 5 4 3 2 1

ISBN 0 431 07622 7

British Library Cataloguing in Publication Data
Oxlade, Chris
The Earth and beyond. – (Science topics)
1. Astromomy – Juvenile literature
I. Title
520

Acknowledgements
The Publishers would like to thank the following for permission to reproduce photographs: Ann Ronan Picture Library pg 21; CFCL/Image Select pg 13; Pix pgs 11, 14; Planet Earth Pictures/Richard Cottle pg 6, /Ed Darack pg 9, /William M Smithey pg 12; Rex Features pgs 8, 15; Science Photo Library/ Chris Madeley pg 23, /NASA pgs 24, 27, 29, /European Space Agency pg 26. Inset artwork for pg 5 by Jonathan Adams.

Cover photograph reproduced with permission of Science Photo Library. Cover shows a composite photograph of the Earth and Moon over a background of a galaxy swirl.

Our thanks to Jane Taylor for her help in the preparation of this edition.

Any words appearing in the text in bold, **like this**, are explained in the Glossary.

Contents

Planet Earth

The Earth was formed about 4,500 million years ago from material left over after the Sun was formed. The Earth will not last forever – in thousands of millions of years it will be destroyed when the Sun dies.

Science essentials

There is hot, molten rock called **magma** under the Earth's surface. Magma sometimes comes to the surface to cause volcanoes.

Runny inside

When the Earth was formed, it was a huge ball of hot rock. Over millions of years, the outer layers cooled to form a thick, hard **crust**. The crust, which we live on, sits on top of a layer of hot rocks, called the **mantle**. In the top part of the mantle, just under the crust, the rock is molten, and flows like treacle. It is called magma. It sometimes comes to the surface, causing volcanoes. In the centre of the Earth is a solid core. Here the temperature is about 4500° C, but the rock does not melt because the **pressure** caused by the weight of the rocks above is so immense.

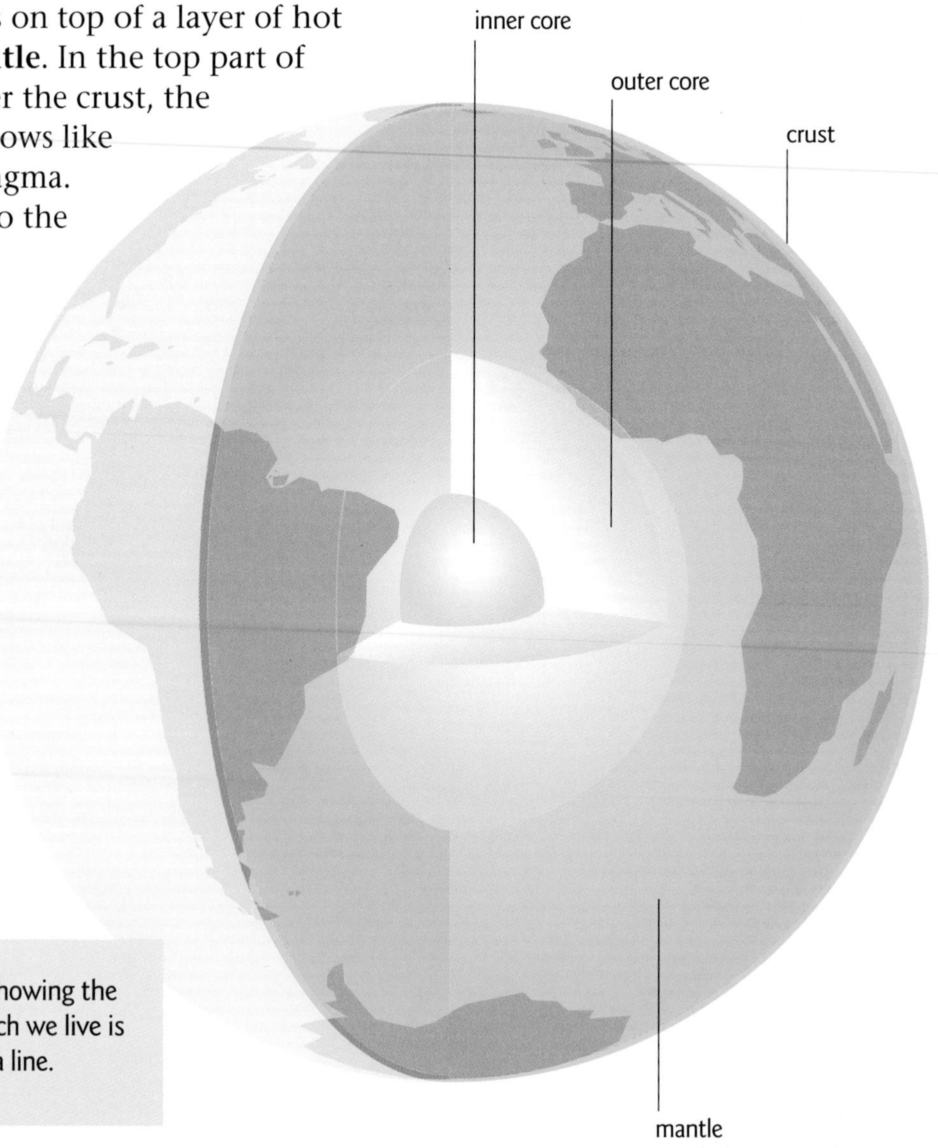

A cross-section of the Earth showing the main layers. The crust on which we live is so thin that it shows only as a line.

Plate movements

The surface of the Earth is changing all the time. The Earth's crust is cracked into many large pieces, which geologists call **tectonic plates**. Magma swirling about under the plates makes them slowly move about. In some places one plate moves away from the next one.

In other places, the plates crash together, or slide and slip past each other. When the plates move in this way they may cause **earthquakes**, volcanic eruptions, and even form new mountains as they crash and force up new land.

Mapping Pangaea

In 1915, before any evidence of tectonic plates existed, a German scientist called Alfred Wegener suggested that all the continents we know today had once made up a single ancient continent, which he called *Pangaea*. Over the last 30 years, geologists have found evidence to support Wegener's theory.

They discovered rocks on different continents that match when the continents are fitted together. More evidence for *Pangaea* comes from **fossils** of ancient plants and animals. For example, the remains of a dinosaur called *Massospondylus* have only been found in Africa and North America. These two areas would have been next to each other in *Pangaea*.

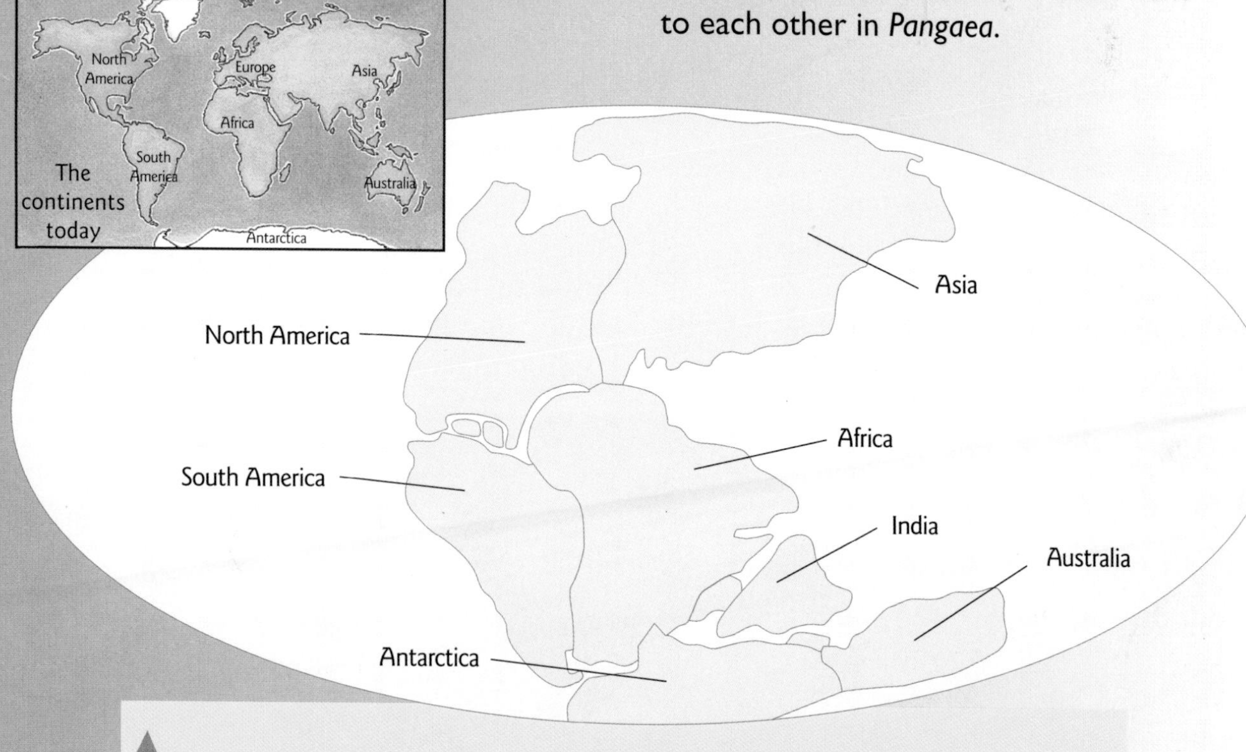

▲ This is what the Earth might have looked like 200 million years ago. The world's continents and islands fit together quite neatly, like a giant jigsaw puzzle.

Rocks and minerals

Rocks are easy to see in mountains and by the sea. In fact, if you dig deep enough anywhere on Earth, you come to the rocks that make up the Earth's **crust** – igneous, sedimentary and metamorphic rocks.

Rocks from magma

When molten **magma** cools, it turns into igneous rock. Some igneous rocks are formed from magma that flows out of **volcanoes** as lava. These rocks are called lava rocks. Other igneous rocks are formed when magma comes close to the surface and cools down.

Igneous rocks are made up of crystals. Lava rocks, such as basalt, cool very quickly in the air, so they have crystals too small to see. Rocks that cool slowly, such as granite, have much larger crystals.

Science essentials

There are four main types of rock:

- **Igneous** rock – formed when **magma** or **lava** is cooled. Rock that cools quickly has smaller **crystals** than rock that cools slowly
- **Sedimentary rock** – usually formed under the sea from layers of **particles** that are compressed together
- **Metamorphic** rock – formed when rock is forced underground where it is changed by high **pressure** and heat
- **Fossils** – formed over millions of years from the remains of animals or plants that become buried in sedimentary rocks.

Building rocks

Sedimentary rock is made when layer upon layer of tiny **particles** settle on the bottom of a sea or lake. Over millions of years, deep layers of particles build up. The particles lower down get squashed by the huge weight of the deposits above. Any water is gradually squeezed out and the particles bind together to form solid rock. Shale, sandstone and limestone are different types of sedimentary rock.

These layers of sandstone were originally flat. They have been folded by movements of the Earth's **tectonic plates**.

Skeletons in the rock

Sometimes, the bodies of dead animals fall to the bottom of rivers, lakes and seas. The soft body rots away, leaving the hard parts, such as bones and shells, which eventually turn into rock as they are buried. These are called fossils. They take millions of years to form. Coal, oil and gas are called **fossil fuels** because they formed from the dead bodies of prehistoric plants and animals as the heat and pressure under the ground changed them into new chemicals. Fossils also provide vital evidence about how life and the world around us has changed over millions of years.

Mixing it up

Sedimentary and igneous rocks are sometimes pushed deep underground by the movements of the Earth's tectonic plates. Deep within the Earth, high temperatures and pressure changes the rock into a new type called metamorphic rock. Lava or magma flowing next to igneous or sedimentary rock can also turn them into metamorphic rock because of the high temperature.

Rocking around

When the Earth was first formed, there were only igneous rocks, formed from magma. These were broken down into tiny fragments (find out how on page 13). These fragments were were washed down rivers to the sea. There, they formed sedimentary rocks.

Some igneous and sedimentary rocks were forced deep underground to become metamorphic rocks. Some metamorphic rock melted to become magma, which returned to the surface through volcanoes to become igneous rock. This process is still continuing. It is called the rock cycle.

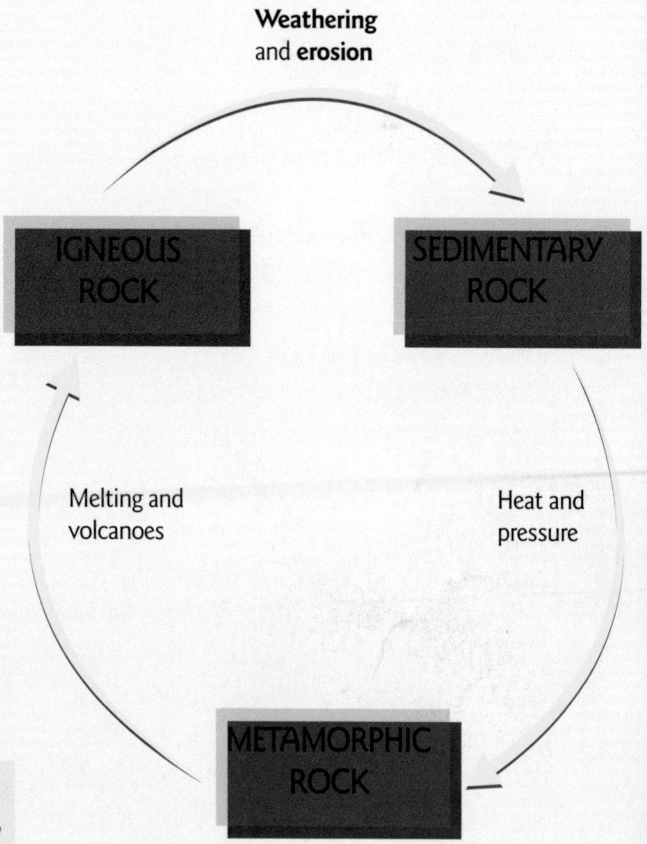

The rock cycle.

Volcanoes and earthquakes

Volcanoes and earthquakes are evidence that there is molten rock beneath the Earth's crust, and that the crust's **tectonic plates** are always on the move. Volcanoes and earthquakes normally happen at the edges of the plates. They can be terrifyingly destructive, wrecking homes, buildings and communications, such as roads and bridges. Volcanoes can also change or destroy animal habitats.

SCIENCE ESSENTIALS

Magma forces its way up through the Earth's **crust** creating **volcanoes**. **Earthquakes** happen because of sudden movements of rocks under the ground.

Magma mountains

Volcanoes happen where magma from inside the Earth forces its way up through cracks or holes in the Earth's crust. Magma that comes out of a volcano is called **lava**. It cools and hardens, making new **igneous** rocks. Over several eruptions, layers of these rocks and of the ash that also comes out of the volcano can build up into cone-shaped volcanic mountains. At the top of the mountain is a lava-filled crater.

Rumblings in the crust

Earthquakes happen where two tectonic plates rub against each other. As the two plates try to move, **friction** stops them from sliding. The plates jolt along, shaking the ground, and causing an earthquake. There are often small earthquakes near volcanoes, but the worst ones happen where two plates collide.

The Earth movements caused by earthquakes and volcanoes can cause huge waves at sea. These are called tsunamis (or sometimes, tidal waves). When they hit the coast, they rear up, becoming tens of metres high, and can sweep many kilometres inland, destroying any buildings or people in their path.

There are many joints between tectonic plates near the islands which make up Japan. This photo shows damage caused by an earthquake in 1995 in which 5,500 people were killed.

Island-building volcanoes

The Hawaiian Islands in the Pacific Ocean are actually the tops of a chain of massive undersea volcanoes. Hawaii itself is built on a volcano called Mauna Loa. Every few years it erupts. Fountains of runny lava shoot from its crater and from cracks on its slopes, and a column of steam and smoke rises upwards. The lava forms red-hot rivers which flow slowly downhill. Many people in Hawaii have lost their homes to the lava flows, but have managed to escape themselves because they received plenty of warning and the lava moves slowly.

Exploding volcanoes

The eruption in 1980 of Mount St Helens in the USA was far more violent. The volcano had been quiet since 1857. The crater was blocked with solid lava. Underneath, though, magma and hot **gases** were building up. There were frequent small earthquakes and then a bulge appeared on the mountainside. Then, one morning, there was a massive explosion. Millions of tonnes of rock were turned to dust and blown nearly 20 kilometres into the sky. An avalanche of hot ash and gas, called a pyroclastic flow, hurtled down from the top of the mountain. When the eruption died down, the mountain was 400 metres lower than before. Explosive volcanoes like Mount St Helens can also cause deadly mudslides as melting ice and snow from the mountain top mix with ash.

Whole forests of trees were flattened, and dozens of forestry workers and campers were killed by the eruption of Mount St Helens volcano in Washington State, USA, in 1980.

The atmosphere

Between the Earth's changing surface and outer space is a thick blanket of air called the atmosphere. It contains nitrogen, oxygen, a varying amount of **water vapour**, and small amounts of other **gases**. These gases are needed for plants and animals to live, to protect us from harmful **rays** from space, and to prevent the Earth's surface from becoming too hot or too cold.

Science essentials

The Earth is surrounded by a blanket of air called the **atmosphere**. The weather happens in the lowest layer of the atmosphere.

Layers of the atmosphere

The air higher in the atmosphere presses down on the air lower down, squashing it into a smaller space. The air is thickest next to the Earth's surface. If you could go vertically upwards, you would find that the air gets thinner and thinner, until eventually there is no air at all. The atmosphere extends for hundreds of kilometres from the Earth, but 99 per cent of the total amount of air in it is in the bottom 50 kilometres.

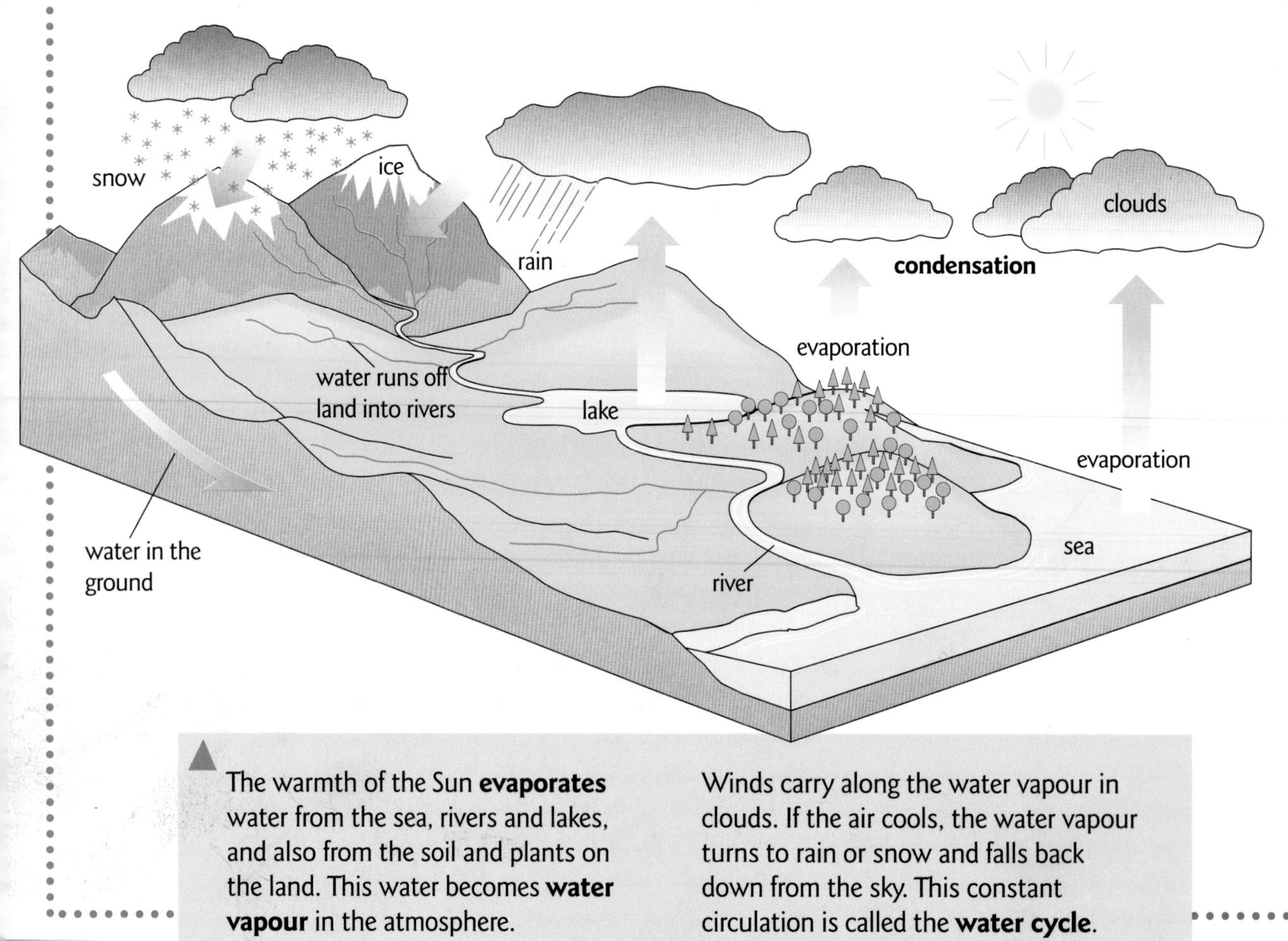

The warmth of the Sun **evaporates** water from the sea, rivers and lakes, and also from the soil and plants on the land. This water becomes **water vapour** in the atmosphere. Winds carry along the water vapour in clouds. If the air cools, the water vapour turns to rain or snow and falls back down from the sky. This constant circulation is called the **water cycle**.

Wind and rain

The weather happens in the bottom layer of the atmosphere, called the troposphere, which is about eleven kilometres thick. Heat from the Sun warms some parts of the Earth more than others. This makes air in one part of the troposphere warmer than the air in other parts. The warmer air expands and rises, and cooler air takes its place moving in from the sides. This moving air is called wind.

Lights in the sky

As well as light and heat, the Sun gives off a stream of sub-atomic **particles** called the **solar wind**, which travels at up to 3 million kilometres per hour towards the Earth. The spinning Earth acts like a huge **bar magnet**. It creates a **magnetic field** which stretches thousands of kilometres out into space. The magnetic field stops the solar wind from hitting the Earth, but it also traps the particles. If the particles collide with gases in the upper atmosphere, they glow, making spectacular patterns called auroras.

The popular name for an aurora display is 'northern lights' or 'southern lights'. The lights can be white, or coloured red, green, yellow and blue. They can make many different kinds of shimmering patterns in the sky.

Shaping the landscape

The Earth's surface is constantly being built up and worn down. New hills and mountains are formed where the Earth's **tectonic plates** crash together and become bent and folded, and by volcanic action. Changes in temperature, and the movement of winds and water in the **atmosphere**, gradually wear away the landscape. Over millions of years, whole mountain ranges are worn away, huge valleys and canyons are carved by rivers, and coastlines are destroyed and created.

Science essentials

Rocks are gradually broken down into small pieces by **weathering**. Weathering is caused by rock being heated and cooled, or by the action of ice. Rocks are also weathered by chemical action.

Physical weathering

Rocks that are heated by the Sun during the day expand (get bigger) slightly because of the heat. At night, they cool down again, and contract (shrink). Over the years, this continuous expansion and contraction makes the rock crumble into tiny fragments. This effect is called weathering.

Another type of physical weathering happens when the rain and snow which fall on mountain tops seeps into cracks in the rocks. In cold weather, it can freeze. Unlike most materials which expand as they get warmer, ice expands as it gets colder. Ice in a crack in a rock expands, widening the crack, and eventually breaking the rock apart.

When water flows over limestone, the limestone dissolves slowly in the water. This is called chemical weathering. Stalactites (which hang down) and stalagmites (which reach upwards) grow when this water drips very slowly, allowing the limestone to form.

Plugs in the landscape

If rocks weather at different speeds, strange shapes can be created in the landscape. After a **volcano** has built up a cone of ash and **lava**, its vent, where the **magma** rises up, can become clogged with hard **igneous** rock. This is called a plug. The volcano then dies. Over millions of years, the cone is weathered away, leaving the plug behind. The plug is often thousands of metres high, with sheer, vertical sides. In some places, plugs are counted as mountains in their own right.

Sugar Loaf Mountain, Rio de Janeiro, Brazil. This mountain is the weathered remains of a volcanic plug.

Rise and shine

Ice weathering can be a major problem for mountaineers. In snow-capped mountains, such as the Alps and Himalayas, the warmth of the Sun melts some of the ice on the mountain tops. If a rock has been loosened by ice weathering, it can fall when the ice melts. Each day, dozens of rocks whizz down the mountainsides. They are all potential killers. Mountaineers often set off on a climb before dawn so they can be off the mountains before the Sun has had time to warm the ice.

Rock in rivers

When it rains, the sand, clay and fragments of rock created by **weathering** can be washed away. They get carried into streams and then on into rivers. In a river or stream this rocky material is called sediment. Eventually, sediment may end up in the sea at the end of a river's journey.

SCIENCE ESSENTIALS

Weathered rocks are carried along by streams and rivers. Streams and rivers **erode** the land as they flow. **Sediment** that reaches the sea forms a **delta**.

Sediment speeds

Very small **particles** of sediment are carried along in the water, but larger particles bump along the river bed. Fast-flowing mountain rivers can move quite large rocks, especially when they are 'in flood'. Further down the rivers, the water slows down, and larger particles of sediment begin to settle onto the river bed, forming layers of mud or silt.

▼ A glacier is a slow-moving river of ice. As it flows, it drags rocks along the ground, gouging valleys into the mountains. The rocks are deposited where the ice melts at the end of the glacier.

Sediment at sea

Tiny particles of sediment are carried all the way to the sea. Here, the river water mixes with the slowly moving sea water and the particles settle on the sea bed. The sediment creates new land around the river mouth. This new land is called a delta. Eventually it can form new **sedimentary rocks**.

Erosion

As sediment moves along a stream or river beds, it scrapes against the bed and bank, wearing them away. This process is called **erosion**. Over millions of years, rivers can cut deep valleys by erosion. If you look in a fast-flowing stream or river, you can see how erosion has worn the rocks on the bed smooth. In windy places, small particles of weathered rock, such as grains of sand, are blown in the wind, scraping away at the rock, too.

Living on a delta

The country of Bangladesh, in southern Asia, lies on a huge delta at the end of two great rivers – the Ganges and the Brahmaputra. Over a year, 2 billion tonnes of silt is washed down the rivers. During the annual floods, a fresh layer of silt is deposited across the delta. For the farmers living on the delta, the flooding of the rivers is vital because the minerals in the silt make the land very fertile for growing crops.

But the floods can also be devastating. The land is so low-lying, at most just a few metres above sea level, that a severe flood can destroy thousands of homes and lives. In a delta, the landscape is constantly changing. Because of erosion and sediment, an island one year may not even exist the following year.

Terrible flooding in Bangladesh in 1991. Floods like this are often made worse by cyclones (similar to hurricanes) sweeping in from the sea.

Light and dark

The Earth is a sphere, and it spins around on its **axis**, an imaginary line which runs through the North and South Poles. Each complete turn of the Earth takes 24 hours. The spinning causes day and night, and makes the Sun, Moon and **stars** appear to move across the sky.

Day and night

When light from the Sun shines on one side of the Earth, the places on that side have daytime. The places on the other side are in shadow at this time – they have night-time. As the Earth spins, places which were in sunlight grow dark, and the places which were in shadow move into the light.

The Earth spins from west to east. That means that places on its surface are moving eastwards all the time. It also means that places to the east of where you live move into the light or shadow before you, so their daytime starts before your daytime, and their night-time starts before your night-time. To the west of where you live, daytime and night-time start later.

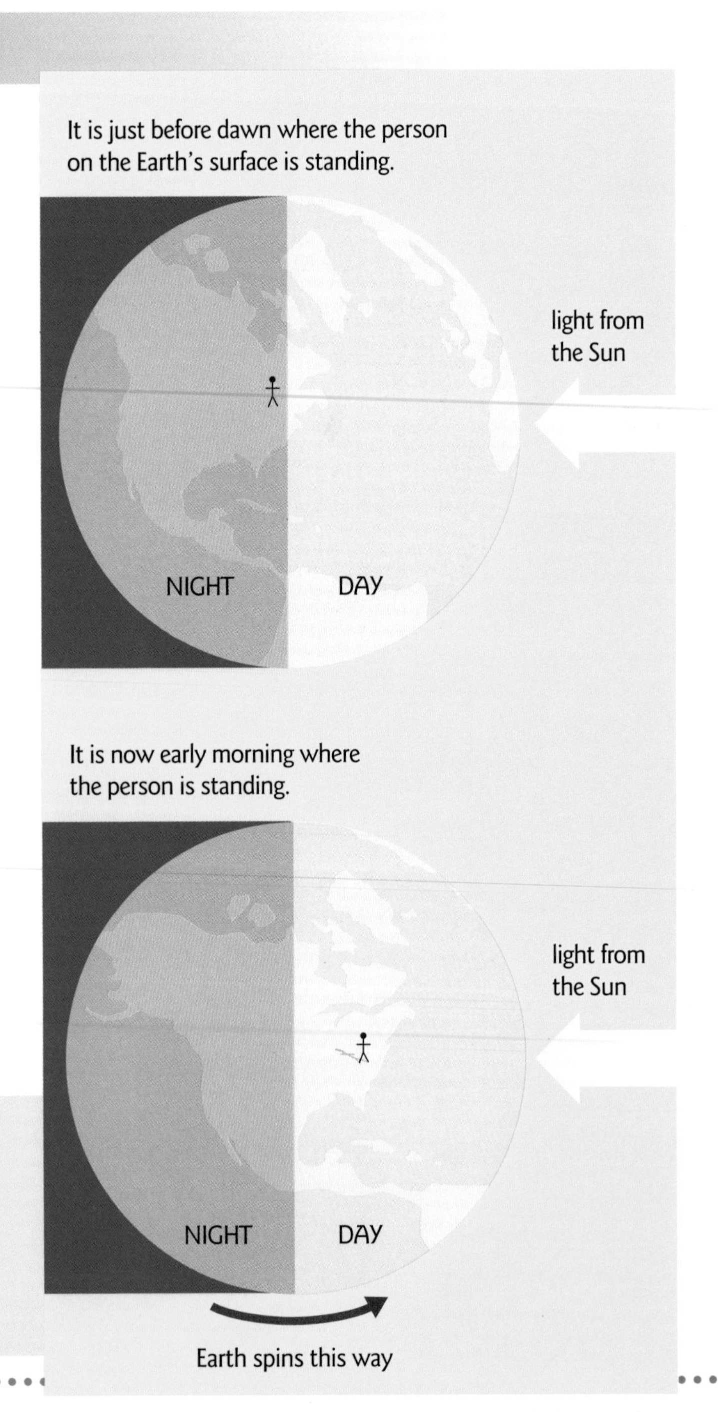

The Sun is shining all the time, but only people in the parts of the world facing the Sun can see it. As the Earth spins, places move from the shadow of night into the light of day in turn.

The moving Sun and stars

The Sun seems to move across the sky. This is because we are watching it from one point on the Earth, which spins round once every 24 hours. Sunrise happens in the east as your part of the Earth moves into the sunlight. Sunset happens in the west when your part of the Earth moves back into the shadow. At night, stars seem to move across the sky for the same reasons.

Science essentials

The Sun and stars appear to move across the sky from east to west because the Earth spins round on its axis. One complete turn of the Earth takes 24 hours.

Time zones

Around 150 years ago, the time in one town could be different to the time in a town just a few kilometres to the east or west. This was because midday was the time when the Sun reached its highest point in the sky. This happened a few minutes later in one town than the other. This did not really matter until train services began to run. Then time had to become standardized. In 1884 delegates from 27 countries met in Washington D.C., USA, and agreed to divide the world into time zones. In each zone, all the clocks are set to the same time. If you move to the next zone, the time is one hour ahead or behind.

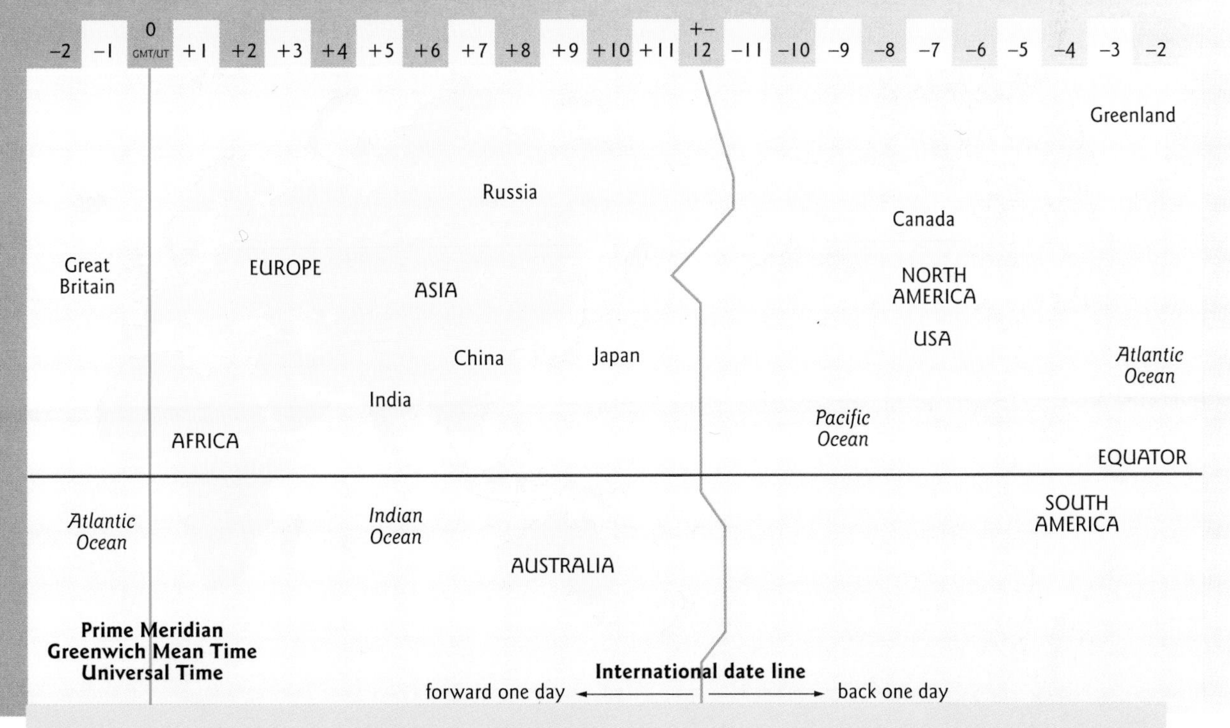

▲ The world is divided into 24 time zones. The numbers on this map show how many hours behind or ahead of Greenwich Mean Time (GMT) or Universal Time (UT) an area is. The orange line is the International Date Line and it marks where one day ends and another begins.

The seasons

Wherever you live on the Earth, the weather will change as the year goes by. This pattern of changes is the same every year. It is called the climate. Away from the warm countries on the **Equator**, most places have four seasons – summer, autumn, winter and spring. These areas have seasons because of the changes in the amount of the Sun's energy that reaches them over the year.

Science essentials

The Earth's **axis** is tilted. As the Earth **orbits** the Sun, the tilt causes the seasons. Different **stars** are visible at different times of year.

A tilted Earth

The Earth's axis is tilted over to one side and stays pointing in the same direction all the time. For half the year, the North Pole is tilted away from the Sun. For the other half, the North Pole is tilted towards the Sun.

When the North Pole is tilted towards the Sun, the northern **hemisphere** gets more sunshine than the southern hemisphere. In the northern hemisphere, the days last longer than the nights. In the southern hemisphere, the nights last longer than the days. This gives spring and summer in the northern hemisphere and autumn and winter in the southern hemisphere.

When the South Pole is tilted towards the Sun, the situation is reversed.

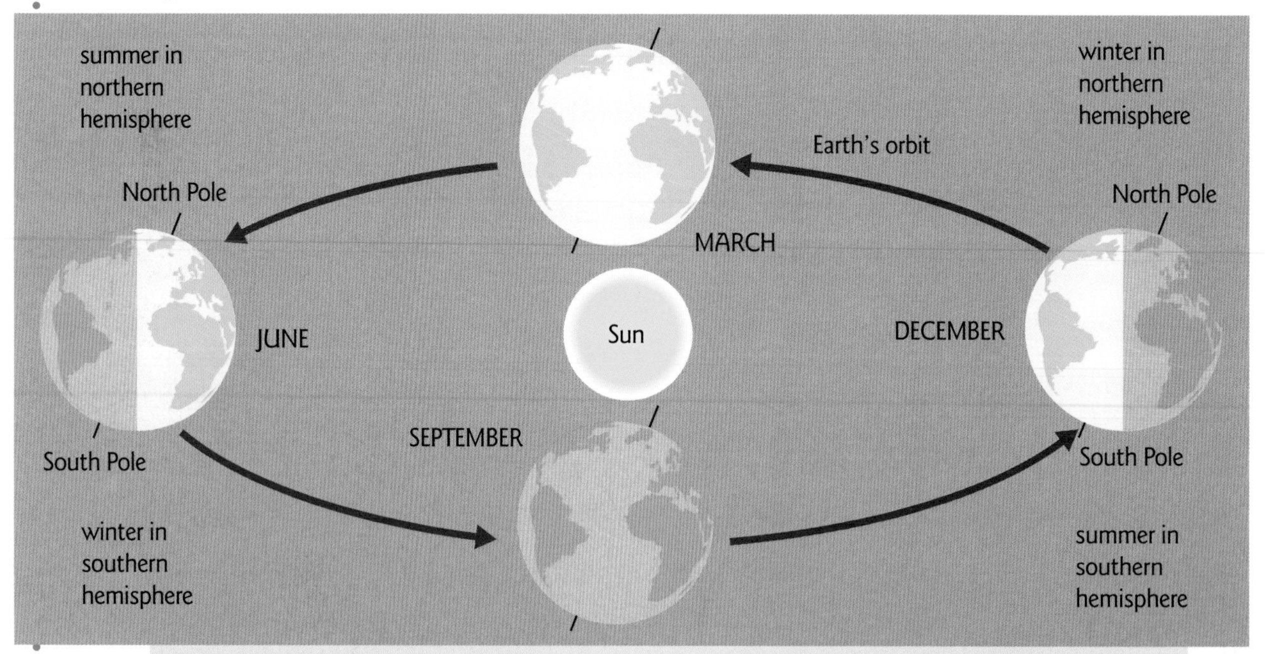

The seasons change according to which part of the Earth is tilted towards or away from the Sun. In June the North Pole is tilted towards the Sun, so it is summer in the northern hemisphere. In December the South Pole is tilted towards the Sun, so it is summer in the southern hemisphere.

Seeing different stars

During the day, you cannot see the stars because the tiny amount of light coming from them is overpowered by the Sun. At night, we can see the stars because we are on the opposite side of the Earth to the Sun. As the Earth moves from one side of the Sun to the other during the year, you can see different stars at night.

▼ These pictures show what the night sky looks like when you look north from the northern hemisphere in midwinter and midsummer. The Pole Star is in line with the Earth's axis, and the other stars appear to spin round it. The patterns formed by groups of stars are called constellations.

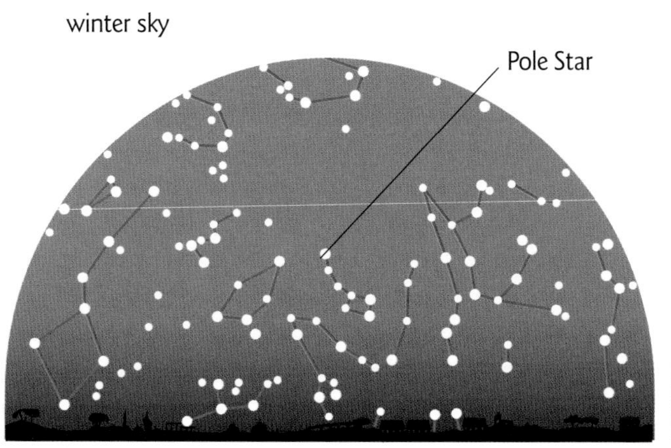

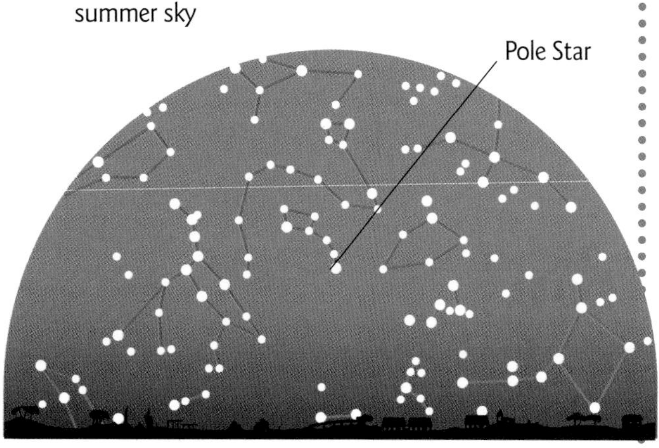

Happy or SAD?

Most of us feel a bit miserable in the winter. The weather is cold, the evenings are dark, and there is not much sunshine to cheer us up. We begin to feel happier when spring arrives and we can feel the warmth of the Sun again.

Some people suffer from worse 'winter blues' than the rest of us. They get a recognized illness called seasonal affective disorder (SAD). They become very depressed in the gloomy days and long nights of winter. Some take medication or even go into hospital.

Scientists think that SAD is caused by a **hormone** in the body called melatonin. It is produced by the body when it is dark, but not when it is bright. If too much melatonin is produced, it makes the person slow and sleepy.

In winter, people who suffer from SAD can be helped if they sit in front of a lamp that creates artificial sunlight for an hour or so every morning and evening to reduce the amount of melatonin in their bodies.

The Sun and the stars

The word 'solar' means 'of the Sun'. The **Solar System** is made up of the Sun, the Earth, and the eight other **planets** that move around the Sun. It also includes the Moon, which moves around the Earth, and the **moons** that move around the other planets.

The path of a planet around the Sun, or of a moon around a planet, is called an **orbit**. The planets and moons, and most other objects in the Solar System, orbit in a regular way, so that we can predict their movements.

SCIENCE ESSENTIALS

The Earth is one of a group of planets that orbit the Sun. The Sun and planets and their moons make up the Solar System. The Sun is the source of light in the Solar System.

Our local star

The Sun is one of many billions of **stars** in a huge group called a **galaxy**. Our galaxy is called the Milky Way. The Sun looks much bigger and brighter than the other stars because it is closer to the Earth. Some stars are millions of times bigger and brighter than the Sun, but they are so far away, they look like tiny pinpoints of light.

The Sun gives off huge quantities of light and heat energy. Only a tiny amount of this reaches the Earth, but without it, plants could not grow, the weather would not happen, and ocean currents could not flow. The Sun is the only object in the Solar System that makes its own light. We can only see the other planets and moons because light from the Sun reflects off them.

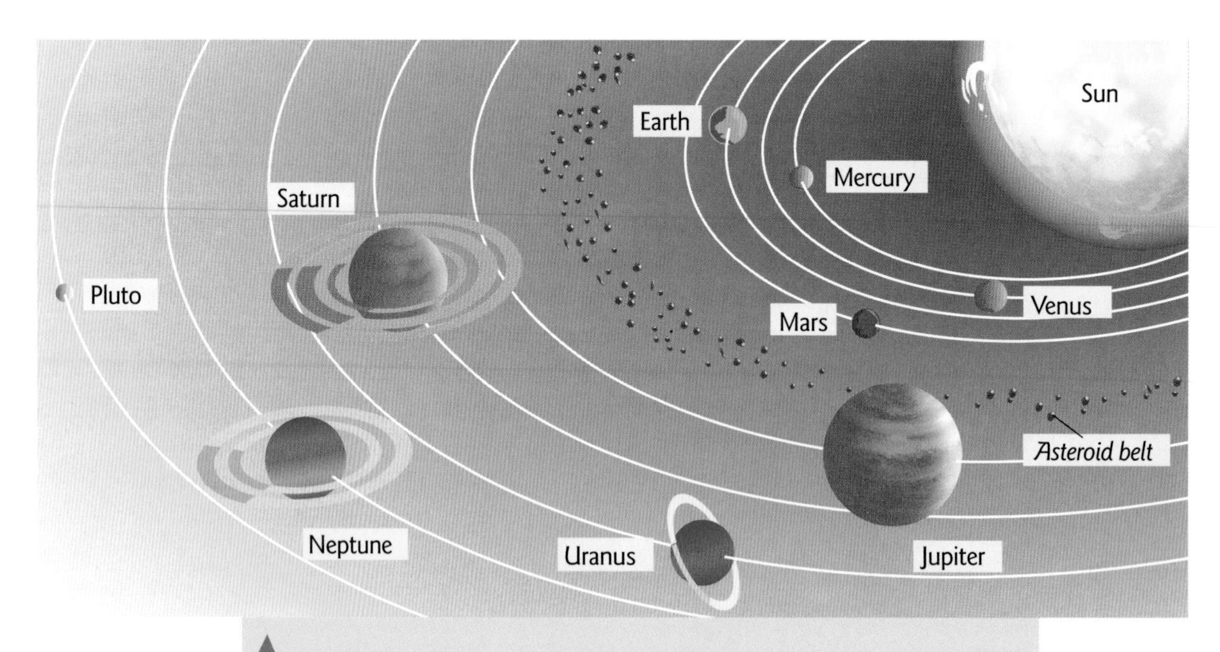

▲ The positions of the planets in our Solar System. This diagram is not to scale – the planets are very tiny compared to the distances between them.

The Earth at the centre

Until the 17th century, when workable telescopes became available, most people, including astronomers, were convinced that the Earth lay at the centre of the Solar System and the Universe. They thought that the Sun, the other planets and the stars all moved around the Earth. After all, this was what appeared to happen.

This theory was put forward by a Greek astronomer, Claudius Ptolemy, in about AD 150. Some astronomers disagreed. One argument was that the stars would have to orbit at an incredibly high speed to travel round the Earth in a day.

They were right, of course, but people liked the idea of the Earth being at the centre of things. In 1530, a Polish priest called Nicolaus Copernicus published a book that claimed that the Sun was at the centre of the Solar System, and the Earth and other planets revolved around it, spinning as they did so. His theory neatly matched the evidence.

Church leaders were furious. They believed that God had created the Earth and placed it at the centre of the Universe. They banned the book, but in the centuries that followed, Copernicus was eventually proved to be right.

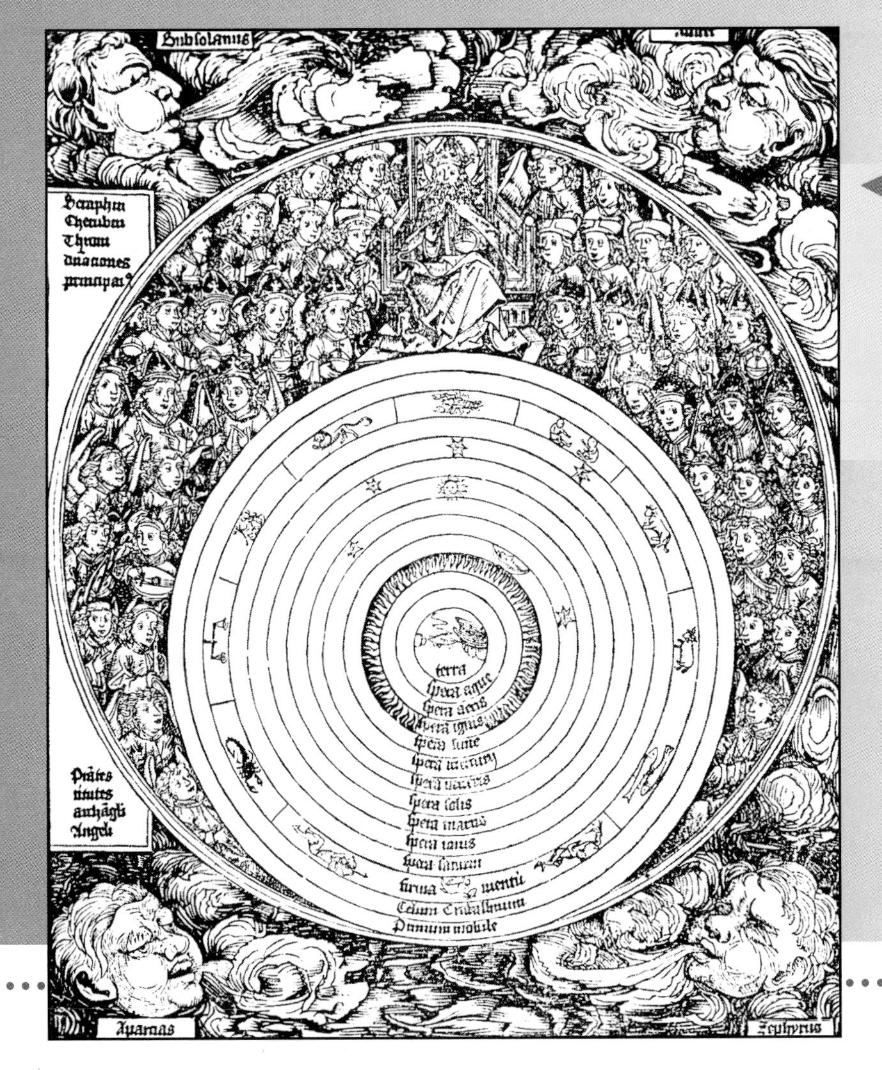

Ptolemy's vision of the Universe. The Earth is at the centre, with the Moon, Sun, stars and the five planets known at the time orbiting around it.

Planets and moons

The four planets nearest the Sun – Mercury, Venus, the Earth and Mars – are planets with rocky surfaces. The next four planets are called **gas** planets. They are really huge balls of liquid, but are called gas planets because they are made from chemicals which are normally found on Earth as gases, such as methane. As you move away from the Sun, each planet takes longer to orbit the Sun. In other words, each planet has a longer year. A year on Mercury, the closest planet to the Sun, lasts 88 Earth days. A year on Pluto, the furthest planet from the Sun, lasts nearly 249 Earth years.

SCIENCE ESSENTIALS

The planets go around the Sun in **orbits** that are nearly circular. We can see the Moon, the other **planets** and their **moons** because sunlight reflects off them.

During each month, the Moon seems to change shape from a full circle to a thin crescent. The Sun can only shine on one side of the Moon. As the Moon orbits the Earth, we can see different amounts of this lighted side. Because the Earth blocks the light, the other parts of the Moon look black.

Moons and rings

The Moon orbits the Earth, just as the Earth orbits the Sun. Each orbit takes just over 27 days. Many other planets have moons, too. We know Saturn has eighteen moons – and there may be more. Saturn has beautiful rings around it, which are made up of billions of orbiting chunks of ice.

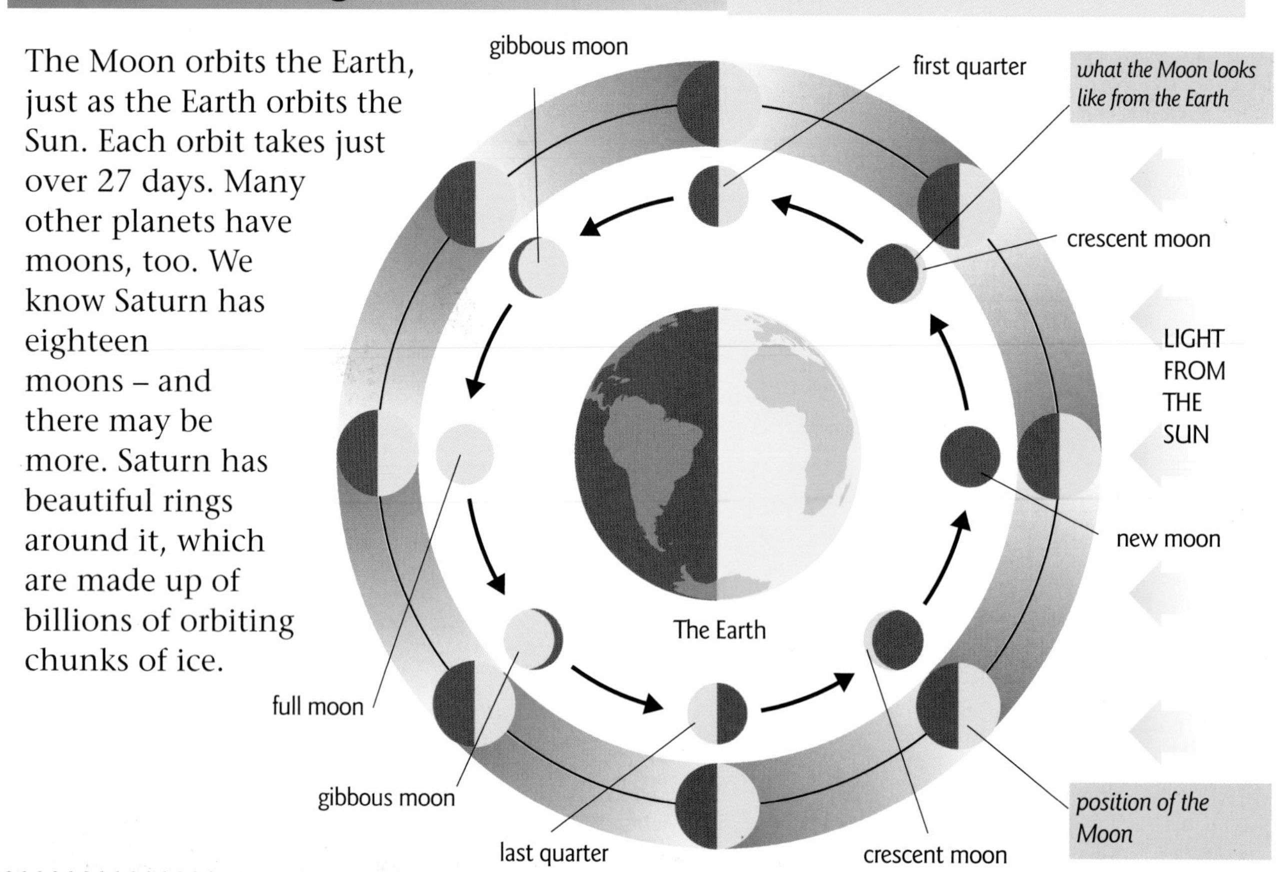

More space objects

Apart from the Moon and planets and their moons, there are many other objects in the **Solar System**. Between the orbits of Mars and Jupiter, millions of chunks of rock called asteroids orbit the Sun. This area is called the asteroid belt (see pg 20). Millions of small lumps of rock, called meteors, whizz through the Solar System. If they hit a planet or moon, they are called meteorites. A shooting star is a streak of light made by a meteor burning up as it travels through the Earth's **atmosphere**.

Comets are lumps of dirty ice and rock. They orbit the Sun in a huge ellipse (oval shape) rather than a circle, spending most of their time far outside the orbits of the planets. When they come near the Sun, the ice begins to turn to gas, and the rock to dust. The **solar wind** blows the gas and dust into a tail millions of kilometres long. The tail disappears again as the comet returns to the outer Solar System.

Collisions

Huge meteorites and comets do occasionally collide with the planets. This has probably happened to the Earth many times over the billions of years since it was formed. The dinosaurs may have become extinct (died out) after a meteorite collision, which would have sent millions of tonnes of dust into the atmosphere, blocking out the Sun for years on end.

Comet Hale-Bopp was one of the brightest comets of the twentieth century. It was discovered in 1995 and could be seen in the night skies from all over the world during 1997.

Gravity at work

Do you know what keeps your feet on the ground? It is the force of **gravity**. Gravity tries to pull objects towards each other, just like a magnet pulls on metal. In fact, gravity tries to pull every object towards every other object. But you only really notice gravity when one or both of the objects is very massive. Gravity pulls everything on Earth downwards. It also keeps the Earth and other **planets** in **orbit** around the Sun.

SCIENCE ESSENTIALS

Gravity is a force that pulls every object towards every other object.
Gravity keeps objects in orbit.
The gravity of the Moon and Sun causes tides on the Earth.

How big is gravity?

The pull of gravity between two objects depends on their **masses** and the distance between them. If the mass of one object increases, the pull increases.

If the distance between them increases, the force decreases. If the distance doubles, the pull actually decreases four times. Once you get a long way from the Earth, the pull is very small indeed. The Moon is smaller than the Earth so it has a weaker gravity pull than the Earth. This would make you feel six times lighter on the Moon, even though you would be exactly the same size.

Gravity and orbits

Planets would not stay in their orbits if there was no gravity. Gravity acts like a tether between the Sun and the planets, stopping the planets flying off into space as they orbit. In fact, without gravity, the **Solar System** would not exist at all. Most astronomers believe that the Solar System was created just under 5 billion years ago when gravity pulled together **gas** and dust **particles** to form the Sun, planets and **moons**.

This astronaut appears to be **weightless**, but he is not. He is simply in the same orbit as the spacecraft that he is in.

Getting into orbit

A spacecraft or **satellite** orbiting the Earth is just like a planet orbiting the Sun. Gravity stops the spacecraft flying away from the Earth, but the spacecraft's movement parallel to the Earth's surface stops it from being pulled down to Earth. The spacecraft is outside the Earth's **atmosphere**. There is no air to slow it down, so it just keeps going without needing an engine.

A spacecraft needs to circle the Earth very fast to stay in orbit. The minimum speed is 28,000 kilometres per hour – any slower and gravity pulls it back down again. When a spacecraft is launched, it starts going upwards, but then begins to tilt over. It keeps **accelerating** and tilting, so that when it flies out of the atmosphere it is moving in a circle at 28,000 kilometres per hour, staying the same distance from the Earth all the time. The engines can be switched off, and it is in orbit. (Find out more about orbits on page 20.)

Gravity and tides

We don't usually notice the small effect that the gravities of the Moon and the Sun have on us because the gravity of the Earth overpowers them. But their effects can be seen in the movement of the tides. As the Earth spins on its **axis** each day, the Moon and the Sun pull at the water, making it slosh about in the oceans. You can see the effects of this if you visit the coast, where the water rises up towards land and falls away from the land twice a day.

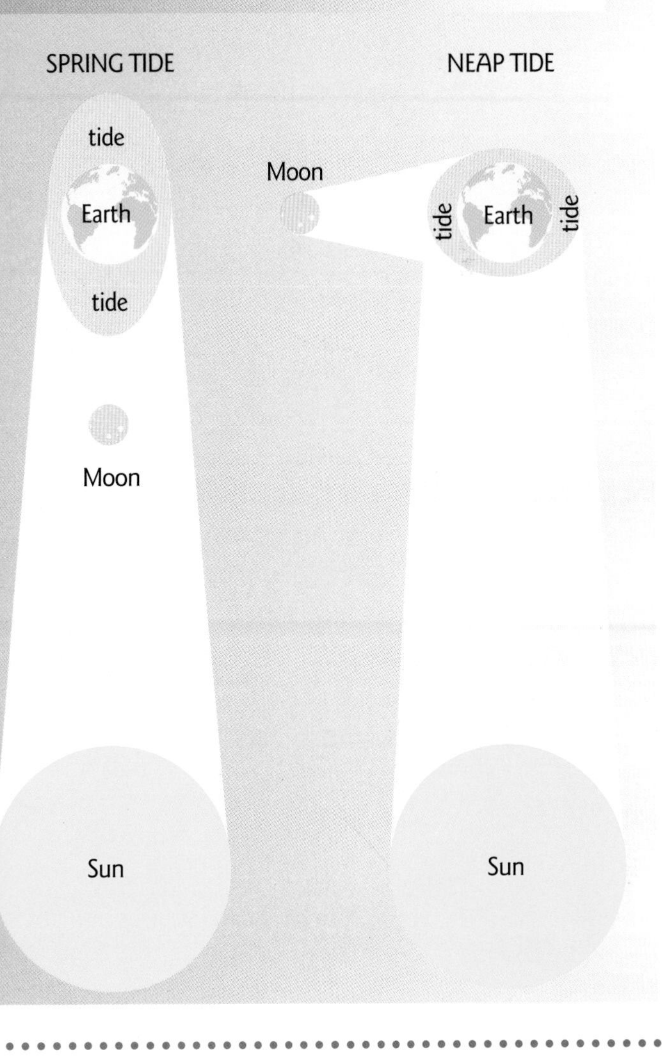

The highest tides (spring tides) happen when the Sun and the Moon are lined up so that they are both pulling in the same direction. When the Moon and the Sun are at right angles, they partly cancel out each other's effect and the tides are lower than average. These are called neap tides.

The Earth from space

You get a really good view of Earth from a spacecraft. You can see whole countries, and even whole continents at once, seen through the clouds swirling above them. You also get a much better view of space than you do from Earth because there is no **atmosphere** in the way.

Science essentials

Artificial **satellites orbit** the Earth. Satellites let us find out more about the Earth. Satellites also give a good view of space.

What is a satellite?

An artificial satellite is a spacecraft that orbits the Earth. There are several hundred satellites in orbit. They work completely automatically for years on end. The majority of satellites around the Earth are for communications. They relay telephone calls, radio messages and television pictures around the Earth. Signals are beamed up to a satellite from the surface. The satellite detects the signals, and beams them back down again. In this way, places that are thousands of kilometres apart can be linked together by a communications satellite.

Looking down at Earth

Satellites that look down on the Earth are called remote-sensing satellites. Weather satellites take pictures that forecasters use to keep track of weather systems, such as hurricanes. They send the pictures down to Earth by radio signals. Some satellites help with scientific research. They detect things such as the heat coming from different parts of the Earth, or the amount of **ozone** in different parts of the atmosphere.

The European Space Agency's ESR-2 satellite being checked before launch. It observes the Earth by bouncing **microwaves** off the surface.

Looking into space

Several satellites help with space research. The Hubble Space Telescope can see much greater detail than the best telescopes on the Earth's surface because it does not have to look through the atmosphere, which blurs the view. The Hubble Space Telescope has taken some extraordinary pictures of distant **galaxies**. Other satellites look out into space, too, searching for **rays** and **particles** coming from the stars and outer galaxies.

Which orbit?

Satellites and other spacecraft are placed in different orbits depending on the jobs they have to do. There are three main types of orbit – a low Earth orbit, a polar orbit and a geostationary orbit. A low Earth orbit is just outside the atmosphere, circling round the Earth above the **Equator**. Here, each complete orbit of the Earth takes about 90 minutes. This is the orbit that the space shuttles and space stations sit in. A polar orbit circles round the Earth so that it passes over the North and South Poles. As the Earth turns and the satellite orbits, the satellite gets a view of the whole Earth over twelve hours. A polar orbit is useful for satellites which survey the Earth, for example for weather forecasting. In a geostationary orbit, a satellite orbits the Earth above the Equator once every 24 hours. Because the Earth spins at the same rate, the satellite stays over the same place on Earth all the time. A geostationary orbit is at nearly 36,000 kilometres above the Earth's surface. Communications satellites use this orbit.

This pod is in low Earth orbit – the same orbit that space stations occupy. It is testing materials to be used in a new space station.

Exploring space

Using large telescopes, we can take photographs of the other **planets** and **moons** in the **Solar System**, but astronomers can find out much more by sending **space probes** to visit these planets and moons. Robot spacecraft have cameras to take photographs and sensors to record other data, such as the temperature of a planet's surface, or the strength of its **magnetic field**. Some probes just fly past other planets; others go into their **orbits**, and some actually land on the surface to investigate their **atmosphere** and their rocks and soils.

SCIENCE ESSENTIALS

A space probe is a spacecraft with no crew which explores other bodies (the Sun, planets, moons and comets) in the Solar System.

A long journey

It is many millions of kilometres to the other planets in the Solar System, and the journey takes many months. Space probes don't have engines of their own. They are launched from Earth by rocket and then simply drift through space. A probe often gains speed by swinging around one planet to reach the next. This is called a gravity assist because the probe uses the **gravity** of the planet to go faster.

A probe's route must be carefully planned so that it arrives at the right point in space to meet the planet it is visiting as the planet moves around its orbit. If the launch is a few days late, the probe could miss its target.

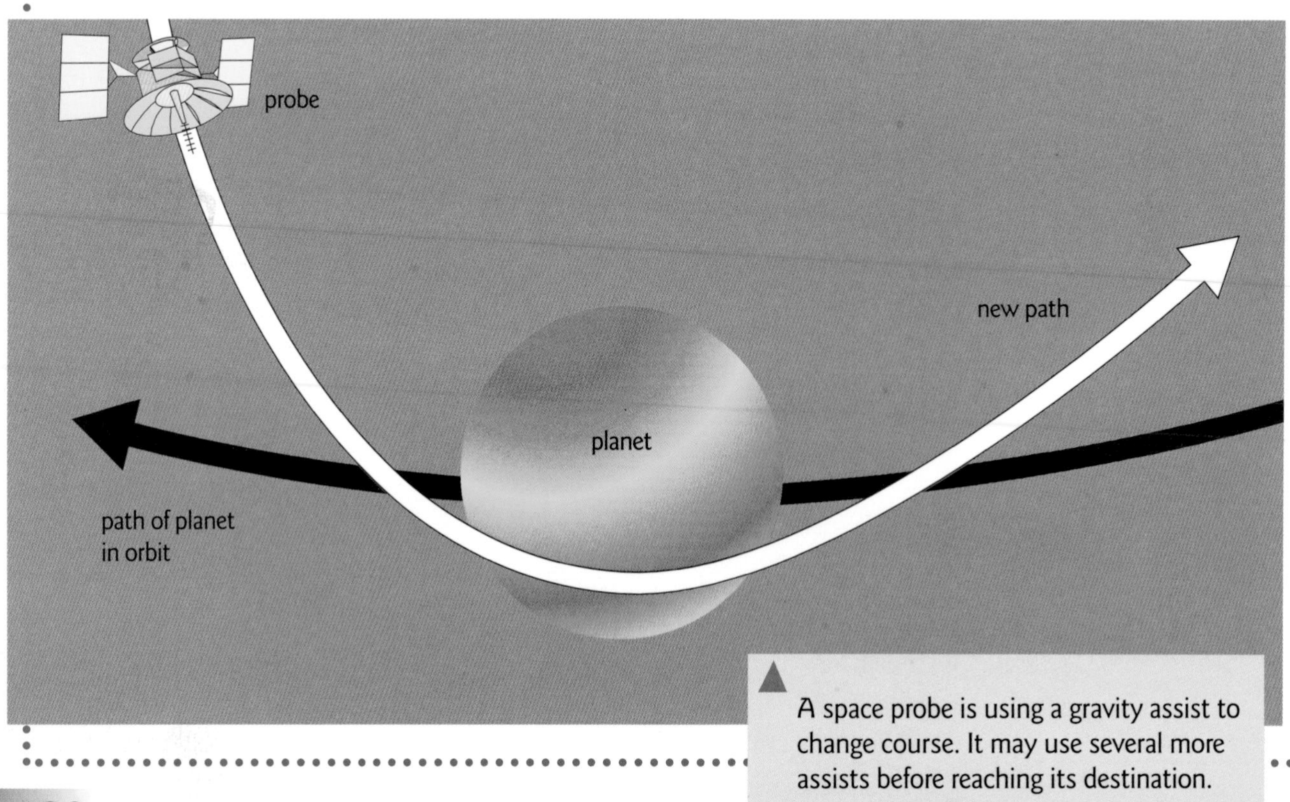

▲ A space probe is using a gravity assist to change course. It may use several more assists before reaching its destination.

Driving on Mars

Mars is our next-door planet, and of all the planets, it looks most like the Earth, with rocks, volcanoes and **ice caps**. In the distant past, rivers flowed on its surface, and it had seas and a thick atmosphere. Scientists want to discover why it is now cold and have sent more probes to Mars than any other planet.

The first successful Mars probe, Mariner 4, flew past in 1965. In July 1997, the probe Pathfinder landed on Mars. Around the outside of Pathfinder were large airbags, which made it look like a huge football. As it landed, the airbags protected the delicate instruments inside from damage. When Pathfinder came to a stop, the airbags deflated and unfolded, like a flower opening its petals. Inside was a small buggy called Sojourner. It was driven by remote control from Earth, down its ramp and across the Martian surface to a nearby rock, which was nicknamed Barnacle Bill.

It took samples of the rock and sent the results back to Earth. Scientists were excited to find that Mars rocks were quite similar to some rocks here. Pathfinder has also taken some amazing pictures of the dusty, red surface of Mars. Another probe, called Global Surveyor, has taken detailed pictures of Mars from orbit.

The robotic solar-powered Sojourner buggy on the surface of Mars in 1997, photographed from the Pathfinder probe. The orange-red colour of the planet's surface is caused by the iron-rich rock that makes up the 'Red Planet'.

Glossary

acceleration speeding up (Deceleration is slowing down)

atmosphere the blanket of gases that surrounds some **planets**. The Earth has an atmosphere of air

axis a straight line about which an object rotates. For example, the axis of a bicycle wheel is a line that goes from one side of the wheel to the other through its centre

bar magnet a magnet the shape of a box or cylinder. It has a magnetic pole (where its magnetic field is concentrated) at each end

condensation the process of **water vapour** in the air turning into liquid water

crust the solid outer layer of the Earth.

crystal a solid substance with a regular pattern of **particles** inside it. Sugar and salt granules are crystals

delta the wide, flat deposit of **sediment** at the mouth of a river. It is created by sediment carried down the river from the land

earthquake a sudden movement in the Earth's **crust** caused by volcanic activity or movements of the **tectonic plates**

Equator the imaginary line round the middle of the Earth that divides the Earth into the northern and southern **hemispheres**

erosion the wearing away of rocks on the Earth's surface by the actions of wind, water and ice, and the seas and oceans

evaporation the process of water turning into **water vapour** in the air

fossil the remains of an animal or plant that has been buried underground and has turned to rock over millions of years

fossil fuel fuels that are formed from the remains of ancient animals and plants which have undergone chemical changes over millions of years. Common fossil fuels are oil, coal and gas

friction a force that tries to stop surfaces sliding against each other

galaxy an enormous group of **stars**, that looks like a cloud in space. Our own galaxy is called the Milky Way

gas one of the three states of matter. The **particles** which make up a gas are widely spaced and move about at high speed

gravity a force that attracts all objects to each other. Gravity between you and the Earth gives you weight

hemisphere half of the Earth, divided by the **Equator** into the northern and southern hemispheres

hormone a chemical made in one part of your body that changes how other parts of your body work

ice cap a thick sheet of ice around the Pole of a **planet**, created by the cold polar climate. Not all planets have ice caps. The Earth has one at each Pole

igneous a type of rock formed from **magma** which cools and hardens

lava the name given to magma that comes out of a volcano and into the air

magma hot molten rock underneath the Earth's **crust**

magnetic field the region around a magnet where the magnet's effect can be felt

mantle the thick layer of hot rocks between the Earth's **crust** and core. The top layer of the mantle is molten **magma**

mass the amount of substance in an object. Mass is measured in kilograms

metamorphic a type of rock formed when **igneous** or **sedimentary rocks** come under great heat or **pressure**

microwaves a type of radio wave that can be aimed accurately and carries a lot of energy

moon a large natural **satellite** of a planet

orbit the path that a **planet** or comet follows as it moves around the Sun, or that a **moon** or artificial **satellite** follows as it moves around a planet

ozone a gas in the upper layers of the Earth's **atmosphere** that helps to stop harmful **rays** from the Sun reaching the Earth's surface

particles microscopically small pieces of a substance, such as molecules in the air

planet a large spherical object that orbits a **star**, for example, the Earth

pressure the amount of force which presses on a certain area

ray a beam of energy that travels in a straight line, such as light

satellite any object (natural or artificial) that **orbits** another object in space

sediment particles of rock and soil that are carried along a river by the flow of the water in the river

sedimentary rock a type of rock formed from **sediment** or the skeletons of sea creatures deposited at the bottom of a sea or ocean

Solar System the Sun, the **planets**, their **moons** and other objects that **orbit** the Sun, such as asteroids and comets

solar wind the stream of sub-atomic **particles** (particles smaller than an atom) that spreads out in all directions from the Sun

space probe a spacecraft that travels from Earth to another part of the **Solar System** to make scientific investigations and send its results back to Earth

star a vast glowing ball of **gas** in space. The Sun is our local star

tectonic plates the series of plates that together make up the Earth's **crust**

volcano a place on the Earth's **crust** where **magma** reaches the surface and cools, often forming a cone-shaped mountain

water cycle the way water from the land **evaporates** into a gas (**water vapour**) in the air and then returns to Earth as rain, hail or snow

water vapour the gaseous form of water, made when water boils or **evaporates**

weathering the wearing away of the rocks of the Earth's **crust** caused by changes in temperature, wind and rain

weightless the feeling of having no weight, caused by being in **orbit** or by being in deep space

Index